B52 049 971 6

KU-303-837

The
Channel Tunnel

◆

MSC

ROTHERHAM PUBLIC LIBRARIES

This book must be returned by the date specified at the time of issue as the Date Due for Return.
The loan may be extended (personally, by post or telephone) for a further period, if the book is not required by another reader, by quoting the above number   LM1 (C)

ROTHERHAM
PUBLIC LIBRARIES

J624.19
P1-4007230
<... 73.3

51587

ROTHERHAM
PUBLIC LIBRAR

SCHOOLS STO

The author and editor wish to thank Christian Antoni and Françoise Delauney of Eurotunnel for their generous help in producing this book.

Picture acknowledgements:
All pictures are from Eurotunnel (photographs by Philippe Demail, G. Liesse, P.-A. Decraene, Q.A. and Phot'R) except for Rouquet/Explorer 6; Giraudon 7, 10 (bottom); Chambre de commerce et d'industrie de Calais 8, 40 (top); Sipa Press 9; Les Editions Albert-René/Goscinny-Uderzo, © 1994 10 (top); Jean-Loup Charmet 11, 12, 15 (top), 17; Roger-Viollet 15 (bottom), 16 (bottom); Sygma 34 (top), 38; Eurostar/SNCF-CAV (D'Angelo) 40 (bottom).

First published in 1994 by Editions Casterman, Belgium.

This translation first published in 1994 by Wayland (Publishers) Ltd 61 Western Road, Hove, East Sussex BN3 1JG, England.

Editor: Katie Roden
Designer: Malcolm Walker
Production controller:
  Janet Slater
Cover illustration:
  Peter Sarson

Copyright © Design and illustration 1994, Casterman, Tournai
© English text 1994 Wayland (Publishers) Ltd

British Library Cataloguing in Publication Data

ISBN 0 7502 1485 6

Timeline

1801   First plan for a fixed cross-Channel link, by the French engineer Albert Mathieu.

1833   Aimé Thomé de Gamond studies different plans before choosing an underground railway tunnel in 1844. He is considered the 'father of the Channel Tunnel'.

1880   First digging works, stopped in 1882 following British opposition to the tunnel. 2,000 m have been dug on the English side, 1,800 m on the French.

1957   Creation of GETM (Groupement franco-anglais d'études du tunnel sous la Manche), which in 1963 proposes building a railway tunnel of two main tunnels and a service tunnel.

1973   Digging starts, from the tunnels dug in the late nineteenth century. The works are abandoned in January 1975 by the British government, for financial reasons.

1981   10–11 September: relaunch of the building plan at the Anglo-French Summit.

1986   The Eurotunnel project is officially approved. 12 February: signing of the Canterbury Treaty.

1987   29 July: approval of the Channel Tunnel Treaty by Margaret Thatcher, British Prime Minister, and François Mitterand, President of France. 15 December: digging starts again.

1990   1 December: historic joining of the service tunnel. Handshake between workmen Philippe Cozette and Graham Fagg.

1993   Work finishes. In December, the builders TML hand over control to Eurotunnel.

1994   Opening of the tunnel. 6 May: official inauguration. July: the Tour de France passes through.

Contents

◆

**Words in bold in the text
can be found in the
Glossary on page 46.**

A Long History

Twelve thousand years ago, our ancestors could walk between England and France. Great Britain has not always been an island. For a long time, an isthmus, a wide strip of land, separated the Channel from the North Sea.

Our planet seems to be still. But we all know that it rotates and also **orbits** the sun. The surface of the earth moves, too. The continents have been slowly shifting over millions of years, causing earthquakes, volcanic activity and tidal waves.

The earth's surface is also changed by **erosion**, such as the gorges and canyons carved by rivers. Seas and oceans also cause great erosion which changes the shape of the shores.

This is how, little by little, the crumbly isthmus was worn down by currents and tides. By 10,000 BC, the Channel linked the Atlantic Ocean to the North Sea, by the gulf which we now call the Pas-de-Calais.

Car ferries go back and forth all day, every day, between Dover and Calais. For a long time, the only way you could make this journey was by sea.

The isthmus broke up very slowly, separating Britain from Europe for ever. The gulf is only 33.5 km wide at its narrowest point between Folkestone and Gris-Nez. The sea is not very deep here – 50 m at most.

Although the Channel is narrow, several hundred ships cross between the Channel and the North Sea every day! The north-ward shipping route follows the French coast, and the southward travels down the British coast. There are special routes for the huge petrol tankers (the 'supertankers'), because their cargoes could harm the environment if they collided with another ship.

The route between Britain and France is just as busy, with ferries and hovercraft sailing back and forth many times a day. Shipping has to be very well controlled, especially as the weather is often misty or windy. For the British, the Channel has been a natural defence against invasions for hundreds of years.

Many people have tried to conquer Britain. In 54 BC, Julius Caesar crossed the Channel to threaten the Britons, who allied with the **Gauls** against him. In the following century, the

The wreck of the supertanker *Braer* in the North Sea in January 1993. Supertankers have their own routes in the Channel corridor, to reduce the risk of collisions.

emperor Claudius made the 'Isle of Britain' part of the Roman Empire. Another emperor, Hadrian, was so afraid of the Scots that he built a stone wall, 110 km long, to protect the north of England. However, it was in the south and east that the great invasions of the fifth century happened, when the island was colonized by the Angles, Jutes and Saxons.

The **Normans** landed in their turn in the eleventh century, led by William the Conqueror. His troops beat those of the Anglo-Saxon King Harold at the Battle of Hastings in 1066.

During the **Hundred Years' War**, the Channel and the North Sea were crossed time and time again. The English controlled Calais for many years, but left France at the end of the war.

▶ Not a bad idea!

England then began to concentrate on founding the many **colonies** that would make it rich. Its fleet, the dreaded Royal Navy, fought for its independence many times. In 1588, the Navy scattered the **Spanish Armada**, which was said to be unbeatable. In the early nineteenth century, it stopped an invasion by Napoleon I of France. Then the great **Industrial Revolution** began and engineers started to dream up plans to link Britain directly to France.

▶ The **Bayeux Tapestry**, in Normandy, France, shows the conquest of England by William the Conqueror.

Under the ground, by sea, by air . . . Napoleon thought of many different ways to get his troops across the Channel.

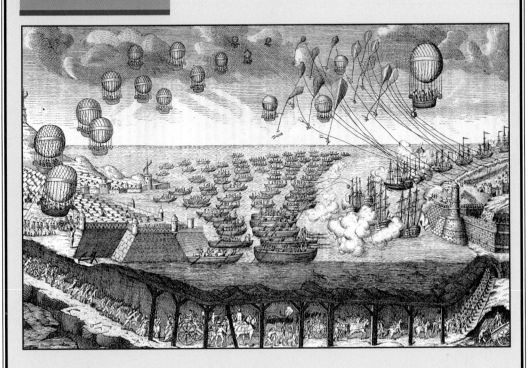

Napoleon I was crowned Emperor of France on 2 December 1804. Like so many other rulers since Julius Caesar, he wanted to cross the Channel and conquer Britain. In several months, 150,000 soldiers, with horses, cannons, equipment and supplies, were gathered on the French coast in hundreds of camps.

The most important camp, Boulogne, housed Napoleon's headquarters and, above all, a fleet of thousands of small boats with flat bottoms, which could run aground on to the beach.

The expedition never set off; in the autumn of 1805 the emperor, nervous of launching a frontal attack, tried to trick the British ships out of the Channel to leave the Pas-de-Calais open for at least ten days. This was a big mistake. The battle of Trafalgar proved, in several hours, the power of the Royal Navy. Napoleon's plans were destroyed. The British still ruled the seas around their island.

The Pioneer Age

Linking Britain to the continent became an obsession in the nineteenth century. Plan after plan was suggested, from the craziest to the most inspired: conveyor belts, sunken or floating tubes, tunnels, dikes, bridges and ferries.

The Industrial Revolution was the age of steam trains and technology. Communication networks were developing at high speed. So were building methods and materials. Roads, bridges and canals, such as the huge Suez and Panama canals, were being opened in many countries. Gustave Eiffel dreamed up daring iron structures like the famous Eiffel Tower in Paris. Tunnels were dug everywhere; the London Underground and the Paris Metro, for example.

The idea of a fixed cross-Channel link became one of the great challenges of this industrial age.

Aimé Thomé de Gamond is known as the 'father' of the Channel Tunnel. However, it was Albert Mathieu who had the first idea for an underground tunnel, in 1801.

The French engineer Aimé Thomé de Gamond (1807-76) is considered the 'father' of the Channel Tunnel. He spent all his energy and money planning a fixed cross-Channel link. He had various ideas, but finally decided on a underground railway tunnel with a very similar route to today's tunnel. To study the sea floor, he made dangerous underwater dives. He weighted his body with bags of pebbles, and filled his ears and nostrils with buttered cotton to protect himself from the water pressure!

Gamond's plans were backed by Emperor Napoleon III of France. In 1867 he won the support of Queen Victoria of Britain, 'in her own name and in the name of all those ladies who suffer from sea-sickness'.

Work started on the tunnel in 1882, when 1,800 m were dug on the French side and 2,000 m on the British. But many British officials were worried that the tunnel would 'destroy the principal defence of this country, upon which it depends the most – isolation.' Work stopped after 1883. About 100 years would go by until digging started again.

Throughout the nineteenth century, many plans had been suggested. In 1856 the English engineer William Austin thought up the basic idea of our modern tunnel: – two railway routes and a service tunnel.

Many people, like the English engineer Henry Mottray, wanted a sunken tube. Gustave Robert suggested a pier with two wide locks to let ships pass through, while Eugène Bruel planned to rebuild the old isthmus. Vérard de Sainte Anne, Charles Boutet and John Winton preferred daring designs for long metal bridges.

Many people liked Gustave Eiffel's plan for long metal bridges to cross the Channel. However, they were impossible to build because of the rocky sea-bed.

The Mont Saint-Gothard tunnel between Italy and Switzerland was built in 1872. By this time, transport links were very advanced.

The crossing of the Channel inspired many nineteenth-century cartoonists. Look at this fantastic sea-going train.

The crossing of the Channel inspired many crazy plans! In 1850, Ferdinand Lemaître revealed his plan for a bridge hanging from fixed balloons. Adrien Huet invented a sea-going steam train which moved on floating barrels, and Dr Lacombe suggested a fantastic underwater train.

By this time, the first regular cross-Channel steamship service had opened in 1820, while Blanchard and Jeffries had first crossed the Channel by hydrogen balloon on 7 January 1785.

At dawn on 25 July 1909, an insect-like buzzing was heard off the coast of Dover. Aboard his little plane, the Frenchman Louis Blériot had just performed a great feat; he had crossed the Channel in 38 minutes.

When he crossed the Channel by plane in 1909, Louis Blériot made aviation history, and started a new age of communication between Britain and the continent.

The First and Second World Wars united Britain and France. Once peace had finally returned, a new era of trade and communications opened up when the **European Community** was formed. In 1957, the plan for an underground railway system was relaunched. Work began in 1973, close to the tunnels dug in 1882. However, an economic crisis hit Britain, and the site was abandoned in 1975 – for ever, it was thought.

The digging work that began in 1882 should have been easy. However, it was abandoned following fierce opposition in British newspapers.

The relaunch did come, though, at the Anglo-French Summit of 1981. Margaret Thatcher, the British Prime Minister, wanted to build the fixed link, as long as the project was paid for by private money. In 1985 several plans were presented to a group of experts, who chose Eurotunnel, an underground rail link.

The Anglo-French Treaty on the tunnel was officially approved on 29 July 1987 by Margaret Thatcher and the President of France, François Mitterand. The building work of the century could begin . . .

In 1967 a high-speed rail link between London and Paris was planned. Work started in 1973, but was abandoned in 1975.

GOD SAVE THE TUNNEL.

FOUR GREAT PLANS

The 'Europont' plan was for a two-storey, six-lane tube, supported by eight pylons, 5,000 m apart.

A huge bridge, made from a tube hanging from eight double pylons, 340 m high and 5 km long – that was the combined bridge and tunnel plan presented by 'Europont-Eurobridge'. The rail link would pass through a sunken tunnel. This project was too expensive.

'Euroroute' was just as daring: a road link with two 500 m-long bridges, joined together by a sunken tunnel, 21 km long. This would be reached by spiral slopes on two artificial islands. The rail link, passing through two sunken tunnels, would be separate.

With its two-track rail tunnel and its road tunnel linked to suspension bridges by spiral ramps, 'Euroroute' was Eurotunnel's closest rival.

'Transmanche Express' consisted of four tunnels, two for rail and two for cars. The cost seemed badly worked out and the ventilation poor.

'Eurotunnel' was based on Thomé de Gamond's original idea. It consisted of two one-way rail tunnels linked every 375 m to a central service tunnel. Vehicles would be transported on special shuttles. This plan had the best design and funding and also protected the marine environment. It was this one that was chosen.

The Building Site of the Century

It would take four years to dig the three galleries of the Channel Tunnel – 152 km of drilling in total. What an amazing challenge!

A huge concrete shaft, 55 m wide and 65 m deep, was dug on the French coast, as near as possible to the sea, at Sangatte. This reached the blue chalk, 47 m under ground, in which the tunnels would be dug. This was impossible at Shakespeare Cliff on the English side, because the coast is edged with cliffs, so a tunnel was built under the sea.

The **geology** of the Channel was important in the planning of the route. The three tunnels had to go through a layer of blue chalk, a soft, fairly solid rock, which is usually found at about 40 m under the **continental shelf** covering the bottom of the sea.

To store the voussoirs at Shakespeare Cliff, a platform was built out into the sea, using the debris from the tunnel.

The layer of blue chalk is found underneath a layer of white and grey chalk. It is quite shallow – 25-35 m – but deep enough for the railway tunnel. This layer sits on a layer of clay, then one of sand.

After many scientific tests on the soil, the route of the tunnel in the chalk could be worked out to the nearest metre. It did not differ much, however, from the one planned at the end of the nineteenth century.

The sub-soil of the Channel is made up of several layers of white, grey and blue chalk resting on a bed of clay.

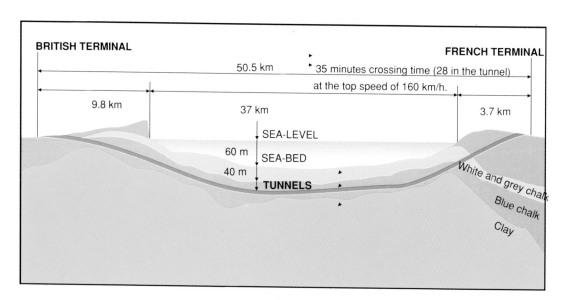

BRITISH TERMINAL

FRENCH TERMINAL

50.5 km

35 minutes crossing time (28 in the tunnel) at the top speed of 160 km/h.

9.8 km

37 km

3.7 km

SEA-LEVEL

60 m SEA-BED

40 m

TUNNELS

White and grey chalk

Blue chalk

Clay

The tunnelling machines were the same width as the tunnel. They hollowed out the clay with powerful blades.

The machines contained a complicated network of pipes and cables, like the insides of a strange mechanical animal.

Advanced technology was needed for such an amazing project. To drill the three tunnels, the engineers used tunnelling machines – gigantic mechanical moles weighing nearly 1,000 tonnes. They were guided by laser beams linked to satellites so they could follow the planned route precisely. Any mistakes could be put right immediately in a control room inside the machine.

The huge, moving factories that followed the tunnelling machines looked like the set of a science-fiction film.

In the cockpit, the tunnelling machines were guided by computer and were kept on course by a laser beam.

It was important that the British and French machines in the same tunnel followed exactly the same path, so that they could meet in the middle. Thanks to electronics, these 'moles' weren't blind at all!

Each tunnelling machine was like an enormous mobile factory, controlled by a team of about 50 people with three pilots. It had a cylinder-shaped engine with an enormous cutting wheel, followed by a 100 to 250 m-long train.

A tunnelling machine was built in the tunnel, and could dig out the earth at 50 mm per minute. It had a spinning drilling head, the same width as the tunnel. This had blades bristling with **tungsten** teeth which ground up the chalk. This was then liquidized and carried to the top of the shaft in big pipes.

The toothed blades of the tunnelling machines chewed up the chalk, turning it to mud. Several million cubic metres of rubble were taken to the surface.

Every 250 m, pressure pipes link the two railway tunnels. They are used to balance the air pressure when trains and shuttles go through.

Conveyor belts brought the **voussoirs** towards the front of the machine. These were ready-made slabs of reinforced concrete which made the framework of the tunnel. Every 1.6 m, the machine laid down some voussoirs and a **keystone**, making the tunnel waterproof. Then it rested on this new concrete ring to continue its underground journey.

HELLO! HOW DO YOU DO?

VERY WELL! ET TOI, COMMENT VAS-TUNNEL SOUS LA MANCHE?

In total, 11 tunnelling machines were used; 6 of them were used under the sea.

On the French side, the chalk lets in more water, so special tunnelling machines, which could move about in a wet environment, were used. They were designed almost like submarines by the Japanese firms (Kawasaki, Marubeni and Mitsubishi) that made them.

In four years, more than 150 km of tunnels were built under the sea!

The tunnelling machines were assembled at the bottom of the service shafts and will never see daylight again. Once they reached their meeting points, some were buried and others were taken to pieces, because the shell created by the voussoirs stopped them from going backwards.

The end of the dig was a fantastic moment for the workers who took part. Everyone wanted to leave his or her own mark!

Digging began in December 1987. Work was slow at first, because the new technology was still not quite right. With 400 people working in shifts, 'Brigitte', or T1, the first French tunnelling machine, took nineteen months and nine days to dig five kilometres.

The entrance to the Sangatte site, with voussoirs ready to be taken down to the tunnelling machines.

However, work got back on schedule. By 1990, about 78 km – more than half the total – had been dug. The first link-up took place at the end of that same year in the service tunnel, 40 m below the sea-bed. On 1 December, at midday, the last piece of blue chalk was cut away. Two workmen, Philippe Cozette from France and Graham Fagg from England, exchanged an historic handshake. From that day on, it was possible to cross the Channel on foot. The two rail tunnels were linked up on 23 May and 28 June; the digging had finished two days before the date set in 1985!

The end of the tunnel at last! The two ends first met in December 1990, in the service tunnel. The dig was completed in June 1991.

AN UNUSUAL FRAMEWORK

The lining for the clay walls of the tunnel was very important. It was a ready-made, cylinder-shaped shell made of extra-strong reinforced concrete. The lining was laid from right inside the tunnelling machine.

Two factories were built, on either side of the Channel, to make the voussoirs. They produced 720,000 reinforced concrete pieces.

So the cylinder could be assembled inside the machine, it was cut into rings weighing up to 9 tonnes each. These rings were also divided up into six parts, called voussoirs. The sixth voussoir, the smallest, formed the keystone of the cylinder, closing the ring when it was put into position. On the French side, the joints between the voussoirs were made waterproof by pressing in a **grout** called 'Neoprene'. This could resist the maximum water pressure of the Channel (6 **bars**, or the pressure caused by 100 m of water).

Waterproof joints were not needed in the drier soil on the English side. The rings were designed so that they could follow the curves of the planned route.

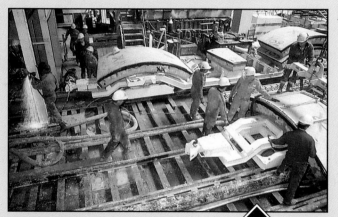

The voussoirs were made on a production line. The concrete was poured into a mould, then dried. The pieces were then stored for about ten days to make them strong enough.

A Guide to the Tunnel

The digging phase was definitely spectacular! But it was only one stage of the project. The tunnel also has to carry travellers and traders from both sides of the Channel as comfortably, safely and quickly as possible.

Let's now look at the three tunnels. The two main tunnels are 7.6 m wide and 22 m apart. They are used by trains, shuttles and the **TGV**. Each tunnel has a one-way track with an overhead cable, and two pavements to make sure the shuttles stay on the rails. The track is made of welded rails on sunken concrete supports. The cooling pipes, fire extinguisher pipes, signalling equipment and cables are attached to the sides of the tunnel.

The trains cross from France to England in the southern tunnel and in the northern tunnel the other way. However, the two tunnels are not completely separate. In fact, they are linked to the service tunnel, every 350 m, by small communication pipes. The service tunnel is 4.8 m wide and is used for maintenance, ventilation and safety.

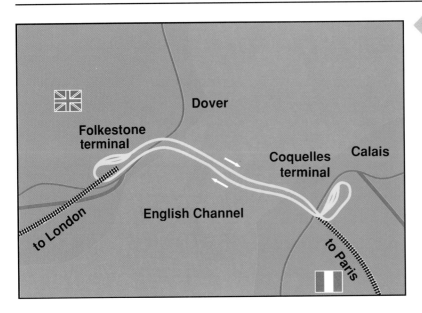

The shuttle system works in a loop between the two terminals. Shuttle trains and direct trains take it in turns to make the crossing.

The communication pipes total 50 km in length. They have 500 fire doors, each weighing 1.5 tonnes.

Pressure pipes directly link the two rail tunnels every 250 m, by passing over the service tunnel. They are 2 m wide, and are used to balance the air pressure. When trains run at high speed through a narrow space they create a huge amount of air resistance. The pressure pipes suck up this air into the other tunnel, which fills the vacuum in front of the train going in the other direction. Clever, isn't it?

The communication pipes are closed off by huge fire doors.

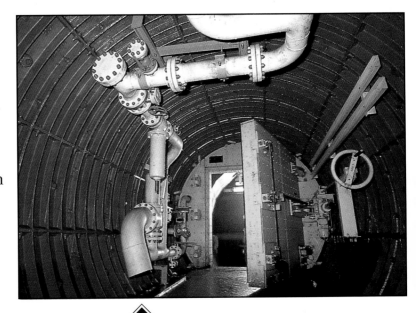

A cross-section of the tunnel. You can see the communication pipes linking the rail tunnels to the service tunnel, and the pressure pipes linking them directly.

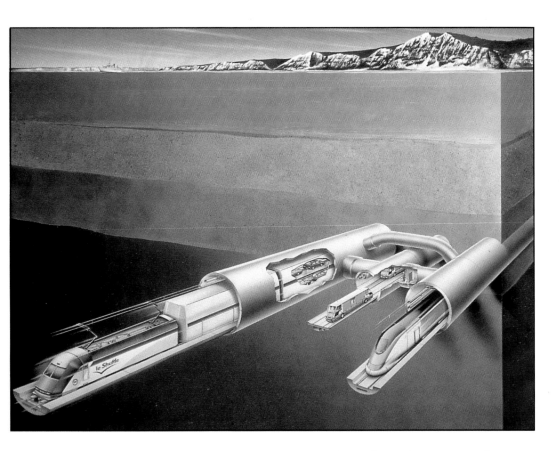

There are also two enormous cross-over chambers between the rail tunnels. These huge galleries, 155 m wide, will allow the trains to change track in the event of fire or mechanical problems, and to be regularly serviced. At normal times, enormous sliding doors weighing 120 tonnes separate the tracks of the two railway tunnels.

The transport system totals 152 km of tunnels. Each tunnel is 50 km long, of which 38 km are under the sea.

The rails are soldered, like all high-speed tracks. They rest on concrete supports.

On the English side, the tunnels continue for over 9 km from Shakespeare Cliff, to emerge at the Castel Hill gate, in the Cheriton area, 60 m above sea-level.

The management of the transport network is based at the control centre at Folkestone. At the centre, monitors show important information about the movements of the trains, the signalling and the state of the ventilation systems, as well as the loading of vehicles on to the shuttles.

The debris from the tunnel made a huge hill on the French coast. Trees will be planted so it matches the environment.

◆ An aerial view of the Coquelles terminal. The terminal area covers 700 hectares, of which 500 are used for the terminal and 200 as a development zone – the 'City of Europe'.

◆ The traffic, signalling and safety systems are controlled from the Folkestone centre.

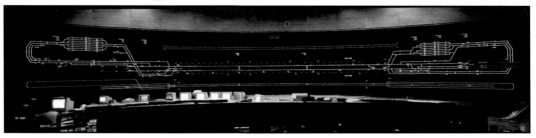

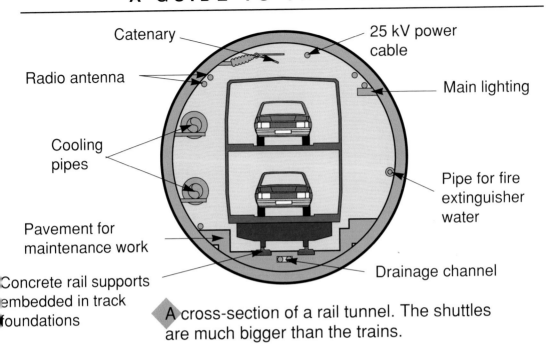

Catenary

25 kV power cable

Radio antenna

Main lighting

Cooling pipes

Pipe for fire extinguisher water

Pavement for maintenance work

Drainage channel

Concrete rail supports embedded in track foundations

A cross-section of a rail tunnel. The shuttles are much bigger than the trains.

Safety is very important in the design of the tunnel. The rail tunnels are separated for daily use so that there is no danger of collision. If there is a fire, a shuttle can keep moving so that the fire can be dealt with in the open air. This is why the carriages must be able to resist flames for 30 minutes. All the passenger shuttles have two engines – among the most powerful in the world – one at the front, the other at the back. If one engine breaks down, the other one can pull the shuttle out of the tunnel. The shuttle can reverse and can be divided into sections. If one section of a shuttle catches fire, it can stay in the tunnel while the other sections are removed in either direction, with all the passengers.

However, it is not possible to transport dangerous substances through the tunnel; petrol or other **flammable** products, **corrosive** chemicals and nuclear waste will still be carried on other routes.

LET'S HOPE THERE ISN'T A STORM!

THE EXPENSIVE CHANNEL TUNNEL

Chosen in 1986 by France and Britain, Eurotunnel runs the tunnel. It was formed by two companies, France-Manche and the Channel Tunnel Group, and will be in charge until 2052.

The building of the tunnel is a great feat, but it has also caused disputes between Eurotunnel, the owner, and TML, the builders.

Eurotunnel gave the planning and building of the tunnel to TML, Transmanche Link, a group of about ten French and English building companies. TML's work took much longer and cost a lot more than the original plan, because the project was so difficult. Originally estimated at about £4 billion, the work and **rolling stock** cost more than three times that amount.

However, the tunnel is still an amazing achievement. More than 12,000 English and French workers were employed on the site between 1986 and 1993, doing a total of 42 million hours of work. At its busiest time, 7,000 people worked at the site. Sadly, there were nine fatal accidents.

A European Challenge

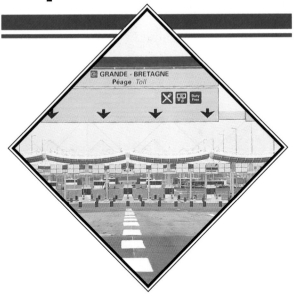

Did you know that 78% of French people have never set foot in Britain? In fact, more than two-thirds of cross-Channel passengers are British. The tunnel should encourage a more even balance.

Nearly 30 million passengers are expected in the tunnel's first few years of operation. With its regular shuttle service, goods trains and high-speed trains, the rail link between Britain and France will eventually be the busiest in the world. With the opening of the tunnel, rail travel will come to rival sea travel. It will encourage the ferry and hovercraft companies, which are mainly British, to improve their equipment and services.

The Channel Tunnel is also an alternative to travelling by air. The direct Paris-London and Brussels-London flight paths are the busiest in Europe and can cause delays at the airports. With a three-hour journey from London to Paris or Brussels on the TGV 'Eurostar', the train rivals the plane for time, cost and reliability.

To compete with the tunnel, the sea companies have modernised their fleets. The new Hoverspeed Seacat crosses the Channel in 35 minutes.

Above all, the opening of the Channel Tunnel will allow the different means of transport to work well together, improving communication between the western European countries. This is especially true for the high-speed train networks.

France leads the rest of Europe with its rapidly expanding high-speed network, the TGV. Opened in 1993, the northern TGV takes only an hour to get from Paris to Lille. Lille is a future departure point for international high-speed links to London, Brussels, Cologne and Amsterdam.

The tunnel will also be good for motorway planning in France and England. The building works have created thousands of jobs for local companies. This has greatly helped the areas which were

The TGV Eurostar will at first take three hours to link Paris to London. However, it will have to reduce its speed through Kent until work finishes on the high speed link.

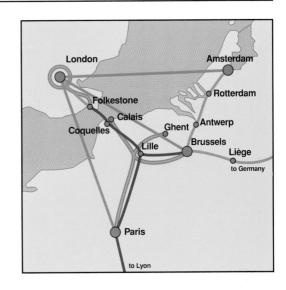

Air links
Motorway links
High-speed rail links
High-speed rail links (planned)

badly hit by the economic crisis, such as Nord-Pas-de-Calais in France. Now the tunnel is finished, many business and leisure centres have been built around the terminals on both coasts, such as the 'City of Europe' in France.

The Channel Tunnel will be a vital link in the northern European communications network.

Having inspired dreams and obsessions for more than two hundred years, the Channel Tunnel has become real. Today, the greatest building project of the late twentieth century has opened the way to a more united Europe. For, above all, it is bringing people closer together.

How to use the tunnel

The Martins have decided to surprise their children Sophie and Christopher: a week-end in Paris, via the Channel Tunnel!

The car journey from London to Folkestone hasn't been very long on the M20 motorway. Mr Martin has already bought his shuttle ticket at a travel agent's. However, you can buy a ticket just as easily by cash or by credit card when you arrive at the ticket offices.

After a quick break in the passenger terminal, with time to visit the **duty-free** shops and to change money, the Martins drive to the boarding point, where **customs** checks are carried out. Then an official guides the cars to the back of a huge, two-storey shuttle, while coaches and caravans are directed to one-storey carriages.

'And how are lorries transported?' asks Christopher. 'They have their own boarding area,' explains the official. 'They are loaded on to open-top wagons, while the drivers travel in a carriage at the front. One single shuttle can carry 28 heavy-goods lorries in one go.

Cars drive on to the back of huge, two-storey shuttles. Passengers stay in their cars during the crossing.

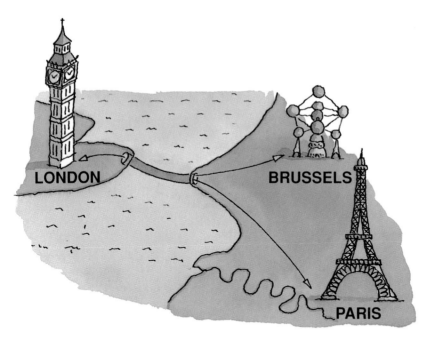

The passenger shuttles carry up to 120 cars at over 140 km/h. And they run all day, every day . . .'

Sophie, who has been carefully studying a leaflet, knows that the shuttles take 35 minutes to get from one terminal to another. Christopher has started his stopwatch to time the crossing, from boarding right up to leaving the system. '55 minutes! Amazing!'

But the journey has only just begun. The Martins' car leaves Coquelles and gets on to the motorway. Have a good holiday!

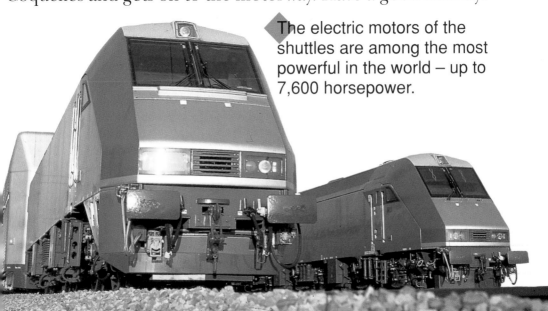

◆ The electric motors of the shuttles are among the most powerful in the world – up to 7,600 horsepower.

On the right track . . .

The Eurotunnel fleet is made up of 38 electric engines, 270 cargo wagons, with 9 passenger carriages, and 252 tourist carriages, as well as the maintenance vehicles.

The shuttle engines pull trains which are up to 800 m long and weigh up to 2,400 tonnes, at a speed of 140 km/h. The shuttles always have two engines, one at either end, which make about 20 crossings a day. The engines get their power from overhead cables of 25,000 volts, and reach a maximum power of 5.6 MW.

DIRECT TRAINS

The tunnels are also used by national railways. This allows services between the main British and European towns. Up to 40 direct trains a day run in either direction. The high-speed Eurostar train takes three hours to get from London to Paris, and three hours ten minutes to Brussels. This journey will be thirty minutes shorter when the British high-speed line is finally laid.

The **gauge** of the rails is the same in Britain as in most European countries, but the British Rail carriages are narrower than the European ones. The carriages of the direct trains have therefore been adapted to run in both Britain and Europe.

The engines that pull the shuttles have two drivers' cabins. The main cabin has a speed-control system.

The shuttles that carry vehicles in the tunnel are much longer and wider than other trains. They can only run in the Eurotunnel system.

NON-STOP TRAFFIC

The Eurotunnel system was designed to allow 20 trains and shuttles per hour to cross in each direction. This can be increased to 30 per hour. The line will soon be the busiest in the world, with about 30 million passengers and 15 million tonnes of cargo in its first few years of service.

The freight shuttles have 14 cargo wagons, two engines and one drivers' carriage. They can carry 28 lorries.

SERVICE TUNNEL VEHICLES

A special transport system runs in the service tunnel. It consists of 24 narrow-gauge **diesel** vehicles. These are steered by an automatic electronic system in the ground, to ensure total safety, but they can also be controlled by hand. They can go in either direction.

RABIES WON'T GET THROUGH

At present, there is no **rabies** in Britain. Some people are worried that animals with rabies will get through the Channel Tunnel and spread the disease.

Eurotunnel has been carefully designed so that foxes and other animals that might carry rabies cannot cross through the tunnel by themselves. However, as with sea or air transport, people who **smuggle** animals can be a risk.

Glossary and Index

Bar The unit by which pressure is measured.

Bayeux Tapestry An eleventh or twelfth-century tapestry in northern France. It is over 70m long, and shows the Norman conquest of England.

Colonies Countries that are controlled by the government of a different country.

Continental shelf The sea-bed surrounding a continent.

Corrosive A word used to describe a substance that burns into other substances, such as acid.

Customs The part of a port or airport where luggage and cargo are checked to make sure nothing is being carried illegally.

Diesel A type of fuel.

Dike A large bank of earth.

Duty-free Goods which are cheaper when sold at airports or docks, because they are not taxed.

Erosion The wearing away of rocks by water, ice or wind.

European Community A western European economic association, formed in 1958.

Flammable A word used to describe something that catches fire easily.

Gauge The distance between the rails on a railway track.

Gaul An ancient region of western Europe, in the area of modern France, Belgium, northern Italy, part of Germany and the southern Netherlands.

Geology The rocks, minerals and soil that are found in the ground.

Grout A type of cement used to fill in the cracks between tiles or bricks.

Hundred Years' War A series of wars between England and France from 1337 until 1453.

Industrial Revolution A period of rapid development in technology and industry in nineteenth-century Britain and Europe.

Keystone The central stone at the top of an arch.

Normans People from Normandy, in northern France.

Orbit To move around another star or planet.

Rabies A serious disease carried by animals.

Rolling stock All the vehicles used on a railway, such as engines, carriages and cargo wagons.

Smuggle To carry something illegally from one country to another.

Spanish Armada A great fleet sent by Philip II of Spain against England in 1588.

TGV 'Train à Grande Vitesse', the French high-speed train network.

Tungsten A very hard metal.

Voussoirs Wedge-shaped stones or bricks, used to build arches.

Index